LIVESTOCK WELFARE AND ETHICAL FARMING

SMILE WELLBECK

GRATITUDE

I want to take a moment to express my gratitude for choosing my book, "Livestock Welfare and Ethical Farming." It means the world to me that you have taken the time to read my work and engage with the topic of animal welfare in farming.

As you know, livestock farming practices have been a highly debated topic in recent years. I wrote this book with the intention of shedding light on the ethical issues in the industry and encouraging a more humane approach that benefits both the animals and the farmers.

I hope that by reading this book, you have gained a better understanding of the importance of ethical farming and animal welfare. My goal is to promote compassion and empathy towards all animals, as well as a more sustainable and equitable farming industry.

Once again, thank you for choosing my book, and I hope you found it insightful and informative. Your support means everything to me, and I look forward to sharing more insights with you in the future.

Copyright © 2024, Smile Wellbeck.

This work and its content are protected under international copyright laws.

No part of this publication may be reproduced, distributed, or transmitted in any form or by any means, including photocopying, recording, or other electronic or mechanical methods, without the prior written permission of the author, except in the case of brief quotations embodied in critical reviews and certain other noncommercial uses permitted by copyright law.

TABLE OF CONTENTS

1. INTRODUCTION TO LIVESTOCK WELFARE AND ETHICAL FARMING
- IMPORTANCE OF ANIMAL WELFARE IN AGRICULTURE
- EVOLUTION OF ETHICAL FARMING PRACTICES

2. ETHICAL CONSIDERATIONS IN LIVESTOCK FARMING
- PRINCIPLES OF ETHICAL FARMING
- ETHICS AND ANIMAL RIGHTS PERSPECTIVES
- THE ROLE OF CULTURAL AND RELIGIOUS PRACTICES

3. ANIMAL WELFARE STANDARDS AND GUIDELINES
- WELFARE STANDARDS FOR DIFFERENT LIVESTOCK SPECIES
- GUIDELINES FOR HOUSING AND SPACE REQUIREMENTS
- NUTRITIONAL REQUIREMENTS AND FEEDING PRACTICES

4. HEALTH AND VETERINARY CARE IN ETHICAL FARMING

- PREVENTIVE HEALTHCARE MEASURES
- DISEASE MANAGEMENT AND TREATMENT PROTOCOLS
- VETERINARY ETHICS AND RESPONSIBILITIES

5. NATURAL BEHAVIOR AND ENRICHMENT PRACTICES
- ENCOURAGING NATURAL BEHAVIORS IN LIVESTOCK
- ENRICHMENT ACTIVITIES FOR PSYCHOLOGICAL WELL-BEING
- IMPORTANCE OF SOCIAL INTERACTION AND HERD DYNAMICS

6. SUSTAINABLE PRACTICES IN LIVESTOCK FARMING
- ENVIRONMENTAL IMPACT OF LIVESTOCK FARMING
- STRATEGIES FOR WASTE MANAGEMENT AND RESOURCE CONSERVATION
- CARBON FOOTPRINT REDUCTION AND RENEWABLE ENERGY USE

7. TRANSPARENCY AND CONSUMER TRUST
- COMMUNICATING ETHICAL PRACTICES TO CONSUMERS
- LABELING AND CERTIFICATION PROGRAMS

- BUILDING TRUST THROUGH ACCOUNTABILITY AND TRACEABILITY

8. INNOVATIONS AND ADVANCES IN ETHICAL FARMING
- TECHNOLOGICAL INNOVATIONS IN LIVESTOCK MANAGEMENT
- RESEARCH AND DEVELOPMENT IN ANIMAL WELFARE SCIENCE
- FUTURE TRENDS AND DIRECTIONS IN ETHICAL FARMING

9. CASE STUDIES AND BEST PRACTICES
- SUCCESSFUL MODELS OF ETHICAL FARMING OPERATIONS
- LESSONS LEARNED FROM GLOBAL PERSPECTIVES
- CHALLENGES AND OPPORTUNITIES FOR IMPROVEMENT

10. CONCLUSION AND FUTURE OUTLOOK
- SUMMARY OF KEY FINDINGS AND INSIGHTS
- RECOMMENDATIONS FOR POLICY MAKERS AND STAKEHOLDERS
- THE PATH FORWARD: ENHANCING LIVESTOCK WELFARE AND ETHICAL FARMING

ABOUT THIS BOOK

Welcome to this guide on 'Livestock Welfare and Ethical Farming'!

This book is designed to provide a comprehensive overview of the concepts of livestock welfare and ethical farming in a language that is easy to understand and relatable. It recognizes the importance of animal welfare and ethical concerns that come with current animal farming practices, and presents a range of practical solutions that can be implemented to reduce animal suffering.

This guidebook will take you through the concepts of animal welfare, including how to identify signs of stress and distress in livestock. You will also learn about the different methods of farming such as indoor and outdoor farming, free-range farming, and organic farming. It will also highlight the benefits and drawbacks of these methods.

Furthermore, this guidebook examines modern farming practices such as the use of antibiotics, growth hormones, and genetic modification, and the impact it has on animal welfare and food products. It will give you insight into the development of livestock welfare and ethical farming regulations and standards across the world.

Ultimately, our hope is that this guidebook will not only provide you with valuable knowledge and understanding of animal welfare and ethical farming, but also inspire you to take action towards a more humane and sustainable way of farming.

We believe that every individual has the power to make informed decisions that have a positive impact on animal welfare, the environment, and human health. With the help of this guidebook, we hope to empower you to make a difference in your own way, no matter how small, towards a better future for our planet.

1. INTRODUCTION TO LIVESTOCK WELFARE AND ETHICAL FARMING

IMPORTANCE OF ANIMAL WELFARE IN AGRICULTURE

In the realm of livestock welfare and ethical farming, the significance of prioritizing animal welfare cannot be overstated. Animals raised in agriculture play a pivotal role in providing food security, livelihoods, and economic sustenance for communities worldwide. However, the manner in which these animals are treated and cared for profoundly impacts not only their well-being but also the sustainability and ethics of the entire agricultural system.

Firstly, ensuring high standards of animal welfare is crucial for maintaining the health and productivity of livestock. Healthy animals are more resilient to diseases and stress, which translates to higher quality products such as milk, meat, and eggs. By investing in the welfare of animals through proper nutrition, housing, and veterinary care, farmers can enhance productivity while minimizing the need for antibiotics and other interventions that may compromise animal and human health.

Moreover, ethical farming practices that prioritize animal welfare contribute to environmental sustainability. Sustainable agriculture aims to minimize negative impacts on the environment, including soil degradation, water pollution, and greenhouse gas emissions. By adopting practices that prioritize animal welfare, such as rotational grazing and integrated crop-livestock systems, farmers can improve soil health, conserve water resources, and reduce their carbon footprint.

Beyond the practical benefits, there is a moral imperative to treat animals with compassion and respect. Animals, sentient beings capable of feeling pain and emotions, deserve to be treated humanely throughout their lives, including during farming practices. Ethical farming practices emphasize the Five Freedoms: freedom from hunger and thirst, freedom from discomfort, freedom from pain, injury, or disease, freedom to express normal behavior, and freedom from fear and distress. By adhering to these principles, farmers not only meet consumer expectations for ethical products but also contribute to a more humane society.

Furthermore, consumers are increasingly demanding transparency and accountability in food production systems. There is a growing preference for products that are produced with respect for animal welfare and ethical standards. Farmers who prioritize animal welfare can differentiate their products in the market, command premium prices, and build trust with consumers who value ethical considerations in their purchasing decisions.

The importance of animal welfare in agriculture extends far beyond the well-being of individual animals. It encompasses environmental sustainability, public health, ethical considerations, and economic viability. Compassion towards animals in agriculture is not just a moral obligation but a pathway towards a more sustainable and humane future for all.

EVOLUTION OF ETHICAL FARMING PRACTICES

In the space of tamed creatures government help and moral developing, a huge improvement is in the works, coordinated by a total moral compass and a creating cognizance of our commitments towards animals and the environment. This change isn't just an example anyway an essential shift towards efficient and thoughtful practices that reexamine our relationship with the food we eat up.

By and large, developing was regularly indivisible from capability and increasing yields, habitually to the disadvantage of animal government help. Regardless, as society's still, little voice has mixed to the circumstance of animals in cultivating, an adjustment of standpoint has occurred. Moral developing presently puts the success of animals at the front, recalling that them as things as well as mindful animals justifying appreciation and care.

Movements in veterinary science and social investigation play played essential parts in this progression. These pieces of information have uncovered old legends and given trial confirmation of animals' capacity for torture, enjoyment, and social bonds. Outfitted with this data, moral farmers are taking on practices that emphasis on animal prosperity and happiness.

Widely inclusive Ways of managing Farming

Moral developing loosens up past animal government help to incorporate sensible agribusiness practices. Farmers are embracing normal methods, rotational brushing, and separated altering systems that further develop soil wealth and lessen reliance on produced inputs. This far reaching approach ensures that farms fill in as one with nature, safeguarding conditions and biodiversity.

Client Consciousness

Perhaps the primary impulse for change has been the upstanding customer. Taught and connected by straightforwardness drives and moral assertions, customers are mentioning food conveyed with dependability. This has helped farmers to embrace moral standards, understanding that ethical practices are morally right as well as financially doable in a market continuously shaped by values.

Challenges and Innovations

Despite progress, challenges persevere. Financial pressures, regulatory frameworks, and social principles can obstruct vast gathering of moral developing practices. Nonetheless, these troubles similarly push advancement. From exactness developing headways to agreeable relationship among farmers and animal government help affiliations, wise fixes are emerging to beat any issues among custom and progress.

The improvement of moral developing practices is a wellspring of motivation for all accomplices in the food structure. Governing bodies ought to initiate plans that help moral standards and give inspirations to sensible agriculture. Buyers ought to continue to zero in on moral examinations in their food choices, driving business area interest for careful developing practices. Farmers, in this manner, ought to make the most of the opportunity to show others how its finished, displaying that moral developing isn't just a moral fundamental yet a pathway to a flexible and prosperous future.

The improvement of moral developing practices tends to a critical shift towards a more others cognizant, legitimate, and extreme food structure. It is an exhibit of our capacity for compassion and stewardship, seeing that the choices today shape the world we leave for individuals later on.

2. ETHICAL CONSIDERATIONS IN LIVESTOCK FARMING

PRINCIPLES OF ETHICAL FARMING

In the domain of domesticated animals cultivating, moral contemplations stand as the bedrock whereupon the whole business ought to flourish. Past simple rural practices, moral cultivating encapsulates a significant obligation to empathetic treatment, ecological stewardship, and supportable practices. It interweaves the government assistance of creatures with the prosperity of the land and individuals who rely upon it. We should dig into the essential rules that characterize moral cultivating and why they are significant for a feasible future.

Regard for Creature Welfare

At the core of moral cultivating lies a profound worship for creature government assistance. Envision a ranch where cows nibble unreservedly on rich fields, chickens peck cheerily at the ground, and pigs root cheerfully in roomy pens. This vision stands out strongly from the confined bounds of production line ranches, where animals are frequently exposed to stuffed conditions, restricted spaces, and upsetting conditions. Moral ranchers focus on the wellbeing and joy of their animals, guaranteeing they approach appropriate sustenance, clean water, and space to show normal ways of behaving.

Natural Responsibility

Moral cultivating stretches out past creature government assistance to include ecological stewardship. Economical practices like rotational touching, treating the soil, and regular bug the board limit the natural impression of cultivating tasks. By supporting the land, moral ranchers save biodiversity, keep up with soil wellbeing, and moderate environmental change influences. They perceive that an agreeable relationship with nature isn't only alluring yet fundamental for people in the future.

Straightforwardness and Accountability

In moral cultivating, straightforwardness rules. Ranchers transparently convey their works on, welcoming buyers to observe firsthand the way that their food is delivered. This straightforwardness encourages trust and responsibility, enabling customers to settle on informed decisions lined up with their qualities. Whether it's through ranch visits, online stages, or direct correspondence, moral ranchers focus on genuineness and respectability in each part of their activities.

Local area Engagement

Moral cultivating is well established in local area values. Ranchers draw in with nearby networks, manufacturing significant associations and adding to the social texture. By supporting nearby economies, giving business open doors, and advancing food security, moral ranchers enhance the networks they serve. They perceive that their activities swell outward, affecting customers as well as neighbors, partners, and people in the future.

Ceaseless Learning and Improvement

Moral cultivating embraces a culture of consistent learning and improvement. Ranchers keep up to date with progressions in animal government assistance science, practical horticulture strategies, and moral norms. They take part in studios, look for certificates, and team up with specialists to improve their insight and abilities. This obligation to long lasting learning guarantees that moral ranchers adjust to advancing difficulties while maintaining the best expectations of care and obligation.

Basically, moral cultivating rises above ordinary farming practices to typify an all encompassing methodology that values respectability, sympathy, and manageability. It addresses a guarantee to making the right decision for creatures, the climate, and society at large. By sticking to these standards, moral ranchers produce nutritious food as well as develop a future where cultivating is inseparable from regard, obligation, and strength.

As shoppers, backers, and stewards of the Earth, we have the ability to help moral cultivating rehearses. By picking items from ranches that focus on animal government assistance, natural supportability, and local area commitment, we can add to a more moral and feasible food framework. Together, we can prepare towards a future where moral cultivating isn't simply a decision yet a common obligation to a superior world.

ETHICS AND ANIMAL RIGHTS PERSPECTIVES

In the core of the discussion encompassing domesticated animals cultivating lies a pivotal thought: the morals and freedoms of the creatures in question. This subject rises above simple monetary or nourishing worries; it dives into the exceptionally upright texture of how we treat conscious creatures that share our planet. As we explore the intricacies of present day farming, it turns out to be progressively basic to assess our practices from a perspective that regards and maintains basic entitlements.

At its center, the moral discussion rotates around the innate worth and government assistance of creatures. Creatures, similar to people, have the ability to feel torment, delight, dread, and satisfaction. This common capacity to encounter feelings frames the premise of our ethical obligations towards them. At the point when we bind animals in squeezed spaces, subject them to distressing conditions, or cause superfluous agony during cultivating rehearses, we disregard these basic freedoms.

Consider the existence of a dairy cow, for instance. Generally, dairy cows are restricted to little nooks for delayed periods, exposed to dreary patterns of pregnancy and draining, frequently prompting physical and mental pressure. Is this treatment ethically legitimate? Advocates for basic entitlements contend that it isn't. They fight that all creatures, no matter what their utility to people, should be treated with pride and sympathy.

Moreover, moral contemplations reach out past the treatment of creatures to envelop more extensive natural and social effects. Concentrated cultivating rehearses, while effective in satisfying worldwide food needs, contribute fundamentally to natural debasement through deforestation, water contamination, and ozone harming substance discharges. These practices compromise creature government assistance as well as undermine biodiversity and worsen environmental change, influencing biological systems and networks around the world.

In tending to these moral difficulties, different points of view arise. Animal government assistance advocates underline the execution of rigid guidelines and others conscious cultivating rehearses that focus on the prosperity of creatures. This incorporates giving adequate space to development, admittance to regular habitats, and limiting pressure during dealing with and transportation.

Then again, defenders of basic entitlements frequently advocate for additional extreme changes, for example, progressing towards plant-based diets or supporting elective protein sources that don't include creature double-dealing. They contend that diminishing or wiping out the utilization of creature items through and through is the most moral position, lining up with standards of peacefulness and regard for every living being.

From a purchaser stance, moral contemplations in domesticated animals cultivating are progressively impacting buying choices. Purchasers are turning out to be more reliable, searching out items named as "accommodatingly raised" or "guaranteed natural" to guarantee that their dietary decisions line up with their moral convictions. This change in customer conduct is driving market influences towards additional supportable and moral practices inside the agrarian business.

The moral components of basic entitlements in domesticated animals cultivating request smart reflection and definitive activity. As stewards of this planet, we have an honest conviction to reexamine our relationship with creatures and the climate. By embracing sympathy, compassion, and supportability in our farming practices, we can endeavor towards a future where the freedoms and government assistance, everything being equal, human and animal the same, are regarded and maintained.

Generally, the moral contemplations in animals cultivating are not just about addressing healthful requirements or monetary productivity; they are tied in with characterizing our identity as a general public and how we decide to coincide with the animals that share our reality. It is through cognizant decisions and aggregate endeavors that we can manufacture a way towards a more sympathetic and morally sound future for all.

THE ROLE OF CULTURAL AND RELIGIOUS PRACTICES

In the unpredictable embroidery of moral contemplations encompassing animals cultivating, social and strict practices weave a significant and frequently profoundly powerful string. These practices, imbued in networks across the globe, shape perspectives towards creatures, their treatment, and the more extensive moral ramifications of our rural practices.

Social standards direct the way that creatures are seen and used inside social orders. For example, in certain societies, steers are adored as consecrated elements representing fruitfulness, overflow, or heavenly association. The meaning of these creatures goes past their financial worth; they epitomize profound and social personalities. Such worship frequently converts into explicit practices pointed toward guaranteeing their prosperity and deferential treatment all through their lives.

Alternately, social practices can likewise legitimize ways of behaving that might be seen as morally argumentative according to different points of view. Practices, for example, custom butcher or conventional hunting might include strategies that outcasts consider brutal or pointless. However, inside the setting of these societies, these practices are much of the time legitimized by well established customs, profound convictions, or saw necessities established in verifiable settings.

Strict lessons further highlight the moral structure of animals cultivating. Numerous strict customs give unequivocal rules on the treatment of creatures, underlining sympathy, stewardship, and regard forever. For instance, Hinduism advances the idea of ahimsa (peacefulness) towards every single living being, affecting dietary decisions and cultivating rehearses among its devotees. Also, Islam and Judaism endorse explicit strategies for creature butcher (halal and fit, separately) pointed toward limiting affliction and guaranteeing the holiness of life.

The crossing point of social and strict practices with current rural practices presents the two difficulties and open doors in exploring moral situations. On one hand, these customs can offer significant experiences into sympathetic treatment and reasonable practices, established in extremely old insight and regard forever. Then again, they may likewise propagate rehearses that contention with contemporary moral guidelines, especially concerning creature government assistance and ecological maintainability.

Tending to these intricacies requires a nuanced approach that regards social and strict variety while advancing moral guidelines that shield creature government assistance and natural honesty. It calls for discourse, instruction, and cooperation between agrarian partners, strict pioneers, social specialists, and moral specialists to figure out some shared interest and cultivate common comprehension.

In pragmatic terms, this implies coordinating social and strict contemplations into more extensive conversations on creature government assistance and supportable farming. It includes investigating imaginative arrangements that accommodate conventional practices with contemporary moral standards, for example, executing compassionate butcher strategies or advancing natural cultivating techniques that line up with strict standards of stewardship and manageability.

Buyer mindfulness and decision assume a significant part in impacting horticultural practices. Progressively, buyers are looking for items that line up with their moral qualities, including contemplations of social and strict awareness. By supporting drives that maintain both moral and social respectability in domesticated animals cultivating, buyers can drive positive change towards additional compassionate and economical farming practices.

Eventually, the job of social and strict practices in moral contemplations of domesticated animals cultivating is perplexing and multi-layered. It requires a reasonable methodology that regards variety, advances exchange, and looks for shared conviction to encourage moral practices that honor both the inborn worth of creatures and the rich embroidery of human social and otherworldly convictions. By embracing these standards, we can endeavor towards a future where farming practices mirror our common obligation to empathy, maintainability, and moral respectability.

3. ANIMAL WELFARE STANDARDS AND GUIDELINES

WELFARE STANDARDS FOR DIFFERENT LIVESTOCK SPECIES

In the domain of creature government assistance, laying out strong norms for the consideration of domesticated animals species isn't simply an ethical basic however an impression of our general public's qualities towards others conscious treatment. Every types of domesticated animals, be it dairy cattle, pigs, poultry, or others, requires custom-made rules that address their special physiological and conduct needs. These norms are not just administrative measures but rather functional structures that guarantee creatures lead lives liberated from pointless torment and pain.

Cattle:

Dairy cattle are social creatures with complex group elements. Government assistance guidelines for dairy cattle stress the arrangement of sufficient room to permit normal ways of behaving like munching and social association. Legitimate lodging conditions that shield them from outrageous climate, adequate admittance to clean water, and a decent eating regimen custom-made to their healthful necessities are principal. Furthermore, taking care of practices should focus on limiting pressure during transportation and veterinary systems.

Pigs:

Pigs are profoundly canny creatures known for their social knowledge and mental capacities. Government assistance rules for pigs center around giving adequate room to development and investigation, as well as ecological enhancement to forestall weariness and energize normal ways of behaving like establishing. Admittance to clean sheet material, fitting ventilation, and appropriate deck that upholds their actual wellbeing are critical perspectives. Torment the board conventions during methodology like emasculation and tail docking are additionally fundamental to limit uneasiness.

Poultry:

Poultry, including chickens and turkeys, require principles that address their remarkable lodging needs and personal conduct standards. Government assistance rules underscore the arrangement of satisfactory room, legitimate ventilation, and lighting conditions that help their wellbeing and prosperity. Admittance to clean water and an eating routine that meets their nourishing prerequisites, including important enhancements, guarantees their development and improvement without compromising government assistance. Besides, taking care of works on during getting and transportation ought to limit pressure and actual damage.

Sheep and Goats:

Sheep and goats are versatile creatures known for their brushing conduct and social designs. Government assistance principles for these species underline admittance to pasture or reasonable open air regions, security from hunters, and asylum from unfavorable weather patterns. Legitimate nourishment, including satisfactory fiber admission, and customary wellbeing checking are basic. Dealing with methods ought to be delicate to try not to cause pointless trepidation or injury.

Consistence and Enforcement:

Carrying out government assistance norms includes joint effort between ranchers, veterinarians, animal researchers, and administrative bodies. Ordinary reviews and reviews guarantee adherence to these norms, giving straightforwardness and responsibility inside the business. Preparing programs for animal controllers and ranchers on prescribed procedures in animal government assistance further support these norms and advance a culture of humane consideration.

The Moral Imperative:

Past legitimate necessities, guaranteeing high government assistance norms for animals species is a moral commitment. Creatures raised for food have the right to carry on with lives liberated from agony, dread, and pain. By embracing these principles, we satisfy our ethical obligations as well as upgrade the personal satisfaction for the creatures that add to our food production network.

Government assistance principles for various domesticated animals species are rules as well as a demonstration of our obligation to sympathetic and moral treatment of creatures.

GUIDELINES FOR HOUSING AND SPACE REQUIREMENTS

Ensuring adequate housing and space for animals is not just a matter of compliance with regulations; it is a fundamental aspect of animal welfare. Proper housing and space contribute significantly to the physical health, mental well-being, and overall quality of life of animals under human care. Whether we're talking about pets, farm animals, or animals in research facilities, the principles remain the same: providing environments that allow them to express natural behaviors, maintain health, and minimize stress.

Why Housing and Space Matter

Imagine being confined in a space where you cannot stretch, move freely, or engage in activities that are natural to you. This scenario applies equally to animals. Insufficient space can lead to physical ailments such as muscle atrophy or obesity in animals unable to exercise adequately. Beyond physical health, cramped conditions can also result in behavioral issues like aggression or stereotypic behaviors (repetitive, seemingly purposeless actions), which indicate stress and frustration.

Key Guidelines
1. Adequate Space: The amount of space required varies depending on the species, size, and behavioral needs of the animals. For example, a large dog needs space to move around and stretch, while a rabbit requires space to hop and explore. Providing sufficient space ensures that animals can exhibit normal behaviors without restriction.

2. *Environmental Enrichment:* Enrichment activities such as toys, perches, hiding spots, and opportunities for social interaction are crucial. These elements stimulate mental activity, prevent boredom, and encourage natural behaviors. For instance, providing scratching posts for cats or puzzles for primates can help maintain their cognitive abilities and reduce stress.

3. *Temperature and Ventilation:* Proper ventilation and temperature control are essential to maintain optimal health. Animals are sensitive to extreme temperatures, which can lead to heat stress or hypothermia if not adequately regulated. Good ventilation helps prevent the buildup of harmful gases and pathogens, ensuring a healthier environment.

4. *Hygiene and Cleanliness:* Cleanliness is vital for preventing diseases and maintaining the well-being of animals. Regular cleaning of enclosures, removal of waste, and disinfection of surfaces are basic but critical practices. This also includes providing clean bedding and fresh water daily, which are essential for health and comfort.

5. *Health Monitoring:* Regular veterinary care and health monitoring are essential to detect and treat illnesses promptly. Early intervention can prevent suffering and ensure that animals receive necessary medical attention to maintain their well-being.

The Ethical Imperative

Beyond legal obligations, there is a moral responsibility to treat animals with respect and compassion. Providing appropriate housing and space is not just about meeting minimum standards but striving for excellence in animal care. It reflects our values as a society that cares about the welfare of all living beings.

Guidelines for housing and space requirements are not mere regulations but essential principles that uphold the dignity and well-being of animals. By adhering to these guidelines, we not only fulfill our ethical responsibilities but also create environments where animals can thrive physically, mentally, and emotionally. Let us strive to implement these guidelines conscientiously, ensuring that every animal under our care lives a life worthy of our respect and compassion.

NUTRITIONAL REQUIREMENTS AND FEEDING PRACTICES

In the realm of animal welfare, few factors are as critical and foundational as proper nutrition and feeding practices. Just as we prioritize our own diets for health and vitality, so too must we consider the nutritional needs of animals under our care. Whether they are pets, livestock, or wildlife in rehabilitation, meeting these requirements not only sustains their physical health but also profoundly impacts their overall well-being.

Understanding Nutritional Requirements:
Every species and even individual animals within species have unique nutritional needs. These requirements are influenced by factors such as age, breed, health status, and activity level. For instance, a young puppy needs different nutrients compared to an adult dog, and a lactating cow requires different feed compared to one that is not producing milk. It is crucial for caretakers to understand these nuances to provide diets that support growth, reproduction, and overall vitality.

Balanced Diets for Optimal Health:

A balanced diet is akin to a symphony of nutrients, each playing a crucial role in the animal's health orchestra. Proteins for muscle development, carbohydrates for energy, fats for insulation and organ protection, vitamins and minerals for metabolic processes—all must be present in appropriate quantities and proportions. Deviations from these optimal balances can lead to malnutrition, obesity, weakened immune systems, and a host of other health issues.

Impact of Feeding Practices:
Beyond the composition of the diet, how and when animals are fed also significantly affects their well-being. Regular feeding schedules help establish routine and reduce stress. Access to clean water is equally vital, as dehydration can quickly become a life-threatening issue. For animals in group settings or pastures, ensuring all individuals have fair access to feed prevents

dominance-related stress and malnutrition among subordinate animals.

Challenges and Innovations:

Meeting nutritional requirements can be challenging, especially in large-scale farming operations or with exotic species in captivity. Innovations in animal nutrition, such as fortified feeds and dietary supplements, have helped bridge these gaps. Additionally, advancements in veterinary science and nutritional research continually refine our understanding of animal dietary needs, allowing for more precise and effective feeding practices.

Ethical Considerations:

At its core, providing adequate nutrition to animals is not just about fulfilling a biological need; it is a moral imperative rooted in our responsibility as caretakers. Just as we advocate for human rights and well-being, we must extend our compassion and diligence to the creatures that depend on us for sustenance. By adhering to high standards of nutritional care, we honor their inherent dignity and contribute to a world where all beings can thrive.

Nutritional requirements and feeding practices form the cornerstone of animal welfare standards. Through informed decision-making, dedication to optimal health outcomes, and respect for individual needs, we can ensure that animals under our care live healthy, fulfilling lives. By advocating for and implementing these practices, we not only uphold ethical standards but also foster a world where compassion and responsibility intersect to benefit all creatures, great and small.

4. HEALTH AND VETERINARY CARE IN ETHICAL FARMING

PREVENTIVE HEALTHCARE MEASURES

In the domain of moral cultivating, where the government assistance of creatures is foremost, preventive medical services estimates assume a vital part in keeping up with the soundness of animals as well as in maintaining moral guidelines and guaranteeing manageability. At its center, preventive medical care epitomizes a proactive methodology that mitigates takes a chance before they raise, advancing animal government assistance as well as protecting the livelihoods of ranchers and the nature of produce.

Figuring out Preventive Healthcare

Preventive medical care in moral cultivating envelops a range of practices focused on prudently tending to potential medical problems in creatures. This incorporates routine immunizations, standard wellbeing screenings, legitimate nourishment, cleanliness conventions, and natural administration rehearses. By carrying out these actions, ranchers can fundamentally decrease the rate of sicknesses, limit the requirement for anti-toxins and other clinical mediations, and at last improve the general prosperity of their animals.

The Significance of Early Intervention

One of the central standards of preventive medical care is early intercession. Recognizing wellbeing worries at their origin permits ranchers to address them, keeping infections from spreading inside the group and staying away from potential financial misfortunes expeditiously. For example, normal veterinary check-ups empower early discovery of conditions, for example, respiratory diseases or parasitic pervasions, which can be quickly overseen through designated medicines.

Advancing Creature Welfare

Moral cultivating puts major areas of strength for an on the others conscious treatment of creatures all through their lives. Preventive medical services measures not just add to keeping up with the actual soundness of animals yet additionally maintain their mental prosperity. By guaranteeing that animals are liberated from superfluous torment and uneasiness, ranchers maintain moral norms and encourage a culture of regard towards the animals under their consideration.

Manageability and Long haul Viability

According to a manageability viewpoint, preventive medical care measures are necessary to the life span and productivity of cultivating tasks. Solid creatures are more useful, requiring less assets for recuperation and yielding better items. This productivity helps the ranch financially as well as lessens the ecological impression related with escalated cultivating rehearses.

Engaging Ranchers Through Knowledge

Integral to the fruitful execution of preventive medical services measures is the strengthening of ranchers with information and assets. Preparing programs and instructive drives assume a urgent part in outfitting ranchers with the abilities to distinguish early indications of disease, direct essential medicines, and carry out preventive procedures really. By cultivating a culture of nonstop learning, ranches can adjust to developing difficulties and maintain the best expectations of animal consideration.

Local area and Partner Engagement

Preventive medical services in moral cultivating stretches out past individual homesteads to envelop more extensive local area and partner commitment. Cooperative endeavors including veterinarians, farming specialists, and administrative bodies guarantee that accepted procedures are shared, principles are maintained, and creative arrangements are created to address arising wellbeing challenges.

Preventive medical services measures are not just a part of moral cultivating; they are its foundation. By focusing on the wellbeing and government assistance of animals through proactive procedures, ranchers maintain moral principles, guarantee supportability, and add to the development of protected and nutritious nourishment for customers around the world. Embracing preventive medical care isn't simply an obligation however a guarantee to the eventual fate of moral cultivating, where sympathy, supportability, and greatness unite to serve all.

DISEASE MANAGEMENT AND TREATMENT PROTOCOLS

In the domain of moral cultivating, the foundation of practical farming falsehoods in the nature of produce as well as in the prosperity of the animals. Sickness the board and therapy conventions assume a vital part in protecting this fragile equilibrium, guaranteeing that our practices are moral as well as compelling in keeping up with wellbeing principles.

Grasping the Importance

Envision a little, family-run ranch settled in the open country, where every animal isn't simply animals yet an individual from the cultivating local area. Here, sickness episodes aren't only dangers to efficiency; they endanger the business and sympathetic stewardship of these creatures. Sickness the executives conventions are hence not simple rules but rather life savers that shield these animals from anguish and the ranchers from monetary strain.

Complete Sickness Prevention

Counteraction is the bedrock of successful illness the executives. Ranches stick to severe biosecurity measures, controlling admittance to the homestead, checking animal wellbeing, and carrying out inoculation programs custom-made to the particular requirements of their domesticated animals. By putting resources into preventive consideration, ranchers relieve the gamble of infections spreading among their animals, consequently lessening the requirement for responsive measures that can be exorbitant and distressing for the two animals and overseers.

Early Discovery and Fast Response

Notwithstanding fastidious preventive measures, illnesses can once in a while break these safeguards. Thus, early identification through cautious observing is significant. Ranchers are prepared to perceive inconspicuous indications of sickness, direct customary wellbeing checks, and immediately disengage any possibly tainted animals. This quick reaction restricts the spread of infection as well as improves the possibilities of fruitful treatment and recuperation.

Comprehensive Treatment Approaches

At the point when treatment is important, moral ranchers focus on the prosperity of their animals. Veterinary consideration centers around restoring the illness as well as on keeping up with creature solace and limiting pressure. Present day veterinary medication offers a scope of treatment choices, from anti-infection agents when fundamental for elective treatments that regard creature government assistance principles. Every treatment choice is made with cautious thought of its effect on the animal's wellbeing and the homestead's general manageability.

Joint effort and Information Sharing

Moral cultivating networks blossom with cooperation and information sharing. Ranchers partake in studios, classes, and industry networks where they trade experiences and best practices in illness the board. This aggregate insight fortifies individual homestead rehearses as well as cultivates a culture of constant improvement in animal wellbeing and government assistance guidelines across the cultivating business.

Straightforwardness and Shopper Trust

Shoppers progressively request straightforwardness in food creation works on, including sickness the executives conventions. Moral ranchers embrace this straightforwardness, sharing their practices transparently with purchasers who esteem capable animal consideration. By building trust through genuineness and respectability, moral ranchers secure their market as well as supporter for the more extensive reception of accommodating cultivating rehearses.

In the domain of moral cultivating, illness the board and treatment conventions are not simply specialized methods; they exemplify a promise to capable stewardship and empathetic consideration. By focusing on preventive measures, quick reactions, and comprehensive medicines, moral ranchers maintain the best expectations of animal government assistance while guaranteeing reasonable and useful cultivating rehearses. This commitment not just shields the wellbeing and prosperity of animals yet in addition improves the networks that depend on moral cultivating for healthy and honestly delivered food.

VETERINARY ETHICS AND RESPONSIBILITIES

In the domain of moral cultivating, where the prosperity of creatures is principal, veterinary morals and obligations assume a crucial part in guaranteeing sympathetic treatment and feasible practices. Veterinarians are not simply medical services suppliers for creatures; they are endowed with the guardianship of their government assistance and the honesty of the farming system.

Veterinarians are limited by a significant obligation to the government assistance of each and every creature under their consideration. This obligation stretches out past simple treatment of diseases or wounds; it includes guaranteeing creatures carry on with a daily existence liberated from pointless misery and with respect. Whether it's a dairy cow, an oven chicken, or a sow in a farrowing case, every creature merits regard and empathetic consideration.

All things considered, circumstances, moral issues frequently emerge. Veterinarians might confront choices with respect to torment the board, fitting lodging conditions, or the moral ramifications of hereditary control. Here, moral structures guide their choices, accentuating helpfulness, non-evil, equity, and independence. These standards guarantee that each choice made lines up with the wellbeing of the animal, the rancher, and the more extensive local area.

Promotion for Moral Practices

Past direct animal consideration, veterinarians act as promoters for moral cultivating rehearses. They team up with ranchers to execute maintainable and compassionate practices that focus on animal government assistance while thinking about the monetary suitability of the homestead. This support includes training, correspondence, and once in a while moving existing standards to advance positive change.

Envision a situation where a veterinarian prompts a poultry rancher on changing from squeezed battery enclosures to improved state lodging. This change works on the government assistance of the chickens as well as improves the nature of their items and the public impression of the ranch. Such changes are moral objectives as well as monetarily useful over the long haul.

Moral Dynamic in real life

Moral dynamic in veterinary consideration requires cautious thought of different partners and moral standards. For example, in instances of sickness episodes, veterinarians should offset general wellbeing worries with the prosperity of impacted creatures. They could suggest quarantine measures, immunization conventions, or even willful extermination when it's to the greatest advantage of forestalling further affliction and transmission.

In addition, moral obligations reach out to straightforwardness and responsibility. Veterinarians should discuss transparently with ranchers and general society about their practices, results of medicines, and any difficulties confronted. This straightforwardness cultivates trust and energizes constant improvement in creature care guidelines.

The Core of Moral Cultivating

Taking everything into account, veterinary morals and obligations structure the core of moral cultivating rehearses. By maintaining trustworthiness, sympathy, and regard for all animals, veterinarians not just guarantee the wellbeing and prosperity of creatures yet in addition add to manageable farming and local area government assistance. Their choices and activities echo through the texture of our food frameworks, forming a future where moral contemplations are fundamental to each part of cultivating.

As we explore the intricacies of current horticulture, let us recall that moral veterinary consideration isn't simply an ethical constraint; it's a foundation of a better, more sympathetic world for animals, ranchers, and customers the same.

5. NATURAL BEHAVIOR AND ENRICHMENT PRACTICES

ENCOURAGING NATURAL BEHAVIORS IN LIVESTOCK

In the domain of domesticated animals the executives, guaranteeing animals display their normal ways of behaving isn't simply a question of animal government assistance; it's a foundation of reasonable and moral cultivating rehearses. Normal ways of behaving incorporate a wide range; from brushing and scavenging to social collaborations and settling impulses; all of which add to the general prosperity and soundness of the creatures. Incorporating these ways of behaving into day to day ranch the executives not just improves the personal satisfaction for domesticated animals yet in addition yields substantial advantages for ranchers and customers the same.

Advancing Touching and Foraging

Permitting animals to brush and rummage unreservedly takes advantage of their inborn ways of behaving of searching out assorted plant species and supplements. This advances better eating regimens as well as imitates their regular natural surroundings, lessening the requirement for exorbitant feed supplements and limiting ecological effect. For example, rotational brushing frameworks further develop soil wellbeing as well as urge creatures to show their normal grouping ways of behaving, cultivating a feeling of local area among the crowd.

Working with Social Interactions

Animals, similar as people, blossom with social connections. Cows have been noticed shaping affectionate bonds inside their crowds, which can lessen feelings of anxiety and improve generally efficiency. Giving adequate space to socialization and trying not to pack guarantees creatures can communicate their normal progressive systems and social elements, prompting a more settled and more satisfied group.

Advancing Natural Stimuli

Presenting ecological improvements like havens, settling materials, and items for investigation invigorates creatures' interest and connects with their critical thinking abilities. For example, giving pigs attaching materials permits them to fulfill their instinctual establishing conduct, forestalling weariness and lessening damaging ways of behaving. This approach works on creature government assistance as well as improves meat quality by decreasing pressure related chemicals.

Supporting Maternal Instincts

Regarding and supporting maternal senses in domesticated animals is urgent for guaranteeing sound posterity and sustaining ways of behaving. Permitting sows to construct homes prior to farrowing and giving more than adequate space to moms and their young advances holding and lessens pressure, prompting better piglets and further developed milk creation.

Teaching and Drawing in Farmers

Empowering regular ways of behaving requires a change in mentality and continuous schooling for ranchers. Preparing projects and assets that accentuate the significance of noticing and understanding animal ways of behaving can engage ranchers to pursue informed choices that benefit both animals and their main concern.

Besides, dividing examples of overcoming adversity and best practices between cultivating networks encourages a culture of nonstop improvement and development.

Uplifting normal ways of behaving in animals isn't just about fulfilling administrative guidelines; it's tied in with developing an amicable and practical connection between animals, ranchers, and buyers. By focusing on these ways of behaving, we improve the prosperity of animals as well as advance ecological stewardship and produce superior grade, morally obtained food. Embracing regular ways of behaving isn't simply great practice; it's a promise to a more brilliant, more reasonable future for horticulture.

ENRICHMENT ACTIVITIES FOR PSYCHOLOGICAL WELL-BEING

In our excursion to explore life's intricacies, cultivating mental prosperity through enhancement exercises becomes a decision, yet a need. Normal way of behaving and improvement rehearses offer a significant pathway to reconnect with ourselves, recuperate from everyday burdens, and prosper genuinely and intellectually.

Drenching in Nature:
Imagine yourself by a quiet lake, the delicate stirring of leaves above, and the mitigating twitter of birds somewhere far off. Nature has an intrinsic capacity to restore our spirits and quiet our psyches. Participating in exercises like climbing, setting up camp, or essentially spending calm minutes in a nursery permits us to loosen up, diminish nervousness, and gain viewpoint. Studies certify that openness to regular habitats brings down cortisol

levels, the pressure chemical, and improves in general state of mind and mental capability.

Embracing Imaginative Expression:
Imagination isn't only for specialists; it's an amazing asset for self-revelation and profound delivery. Whether through painting, composing, playing music, or making, innovative exercises empower us to channel our contemplations and sentiments into substantial structures. These undertakings animate the cerebrum's prize framework, encouraging sensations of achievement and satisfaction. They give an outlet to handling complex feelings and proposition a feeling of satisfaction that rises above day to day schedules.

Supporting Actual Vitality:

The psyche body association is certain with regards to prosperity. Participating in customary actual activity; be it yoga, moving, swimming, or basically strolling; discharges endorphins that elevate state of mind and mitigate pressure. Actual work works on cardiovascular wellbeing as well as improves mental capability and advances better rest designs. It's a characteristic method for helping energy levels and develop strength against life's difficulties.

Developing Care and Meditation:
In our high speed lives, care and reflection act as anchors to the current second. These practices help us to notice our contemplations without judgment, advancing mental clearness and profound equilibrium. By rehearsing care, we foster versatility against pessimistic feelings and develop a more profound identity mindfulness. Research features their viability in decreasing side

effects of wretchedness and nervousness, working on in general mental prosperity.

Encouraging Significant Connections:
 People blossom with social associations and having a place. Taking part in exercises that cultivate significant connections; whether through chipping in, joining clubs, or investing quality energy with friends and family; sustains consistent reassurance and lifts confidence. Solid social ties give a feeling of motivation and satisfaction, buffering against forlornness and advancing generally speaking bliss.

Constant Learning and Growth:
 Scholarly feeling through deep rooted learning improves mental capability and advances self-awareness. Whether through perusing, going to studios, mastering new abilities, or chasing after leisure activities,

consistent learning difficulties our brains and extends our points of view. It cultivates flexibility and strength notwithstanding life's vulnerabilities, engaging us to explore difficulties with more prominent certainty and clearness.

Striking a Balance:

The quintessence of coordinating improvement exercises into our lives lies in tracking down an amicable equilibrium that suits our special necessities. About focusing on exercises feed our brain, body, and soul in the midst of the requests of day to day existence. By integrating these practices step by step and reliably, we leave on an excursion towards all encompassing prosperity and inward satisfaction.

IMPORTANCE OF SOCIAL INTERACTION AND HERD DYNAMICS

In the embroidery of nature, social connection and crowd elements are basic strings that wind around together the existences of endless species, including our own. From the magnificent groups of elephants meandering the African savannah to the clamoring networks of insects teaming up underground, social connection isn't just a component however a foundation of endurance and prosperity.

Building Securities and Fortifying

At its heart, social cooperation encourages bonds that are fundamental for the endurance of people inside a gathering. Consider a crowd of wild ponies exploring the immense fields. Their collaborations are easygoing experiences as well as perplexing trades of correspondence, trust, and participation. Through these collaborations, they lay out pecking orders, care for the youthful, safeguard against hunters, and offer assets — all basic parts of flourishing in their current circumstance.

For people, as well, social cooperation assumes a crucial part. From earliest stages, our advancement depends on collaborations with parental figures and companions. These early encounters shape our interactive abilities, profound strength, and mental capacities. As we mature, kinships and local area ties become indispensable wellsprings of help, having a place, and satisfaction.

Learning and Adaptation

Past simple endurance, social cooperation empowers learning and transformation. In gatherings, people gain from each other, passing on information about food sources, relocation courses, or risk signals. In the set of all animals, this can be seen in the showing ways of behaving of senior elephants directing the youthful or wolves planning chases through unpredictable collaboration.

Essentially, in human social orders, social association drives social development and advancement. It is through joint effort and trade of thoughts that civilizations have flourished, propelling advancements, expressions, and sciences. The elements of interpersonal organizations fuel innovativeness and critical thinking, offering assorted points of view and bits of knowledge that an individual alone might very well won't ever consider.

Profound Prosperity and Mental Health

Also, social association significantly influences close to home prosperity and psychological wellness. Research reliably shows the way that depression and social segregation can adversely affect wellbeing, much the same as chance factors like smoking and corpulence. Alternately, solid social binds connect with expanded life span, strength to stress, and generally speaking joy.

In crowd elements, people get solace and security from their gathering, tracking down strength in solidarity during seasons of difficulty. In human social orders, strong connections support against difficulty, give close to home approval, and deal potential open doors for self-improvement and satisfaction.

Challenges and Resilience

However, the significance of social cooperation stretches out past its advantages. It likewise presents difficulties that empower development and versatility. Compromise, exchange, and compassion are abilities sharpened through exploring complex social elements. The two creatures and people the same figure out how to adjust to contrasting characters, resolve debates, and accommodate clashing requirements; a demonstration of the unique idea of social collaborations.

The meaning of social connection and crowd elements in normal way of behaving and enhancement rehearses couldn't possibly be more significant. It is the establishment whereupon networks flourish, people thrive, and species persevere. From the fields of Africa to clamoring urban communities, the standards of joint effort, compassion, and shared help are general, forming our reality in manners both significant and persevering.

Cultivating conditions that support significant social communications, whether in untamed life protection or local area improvement, we honor the normal request as well as advance our lives immensely. Allow us to embrace the examples of nature's social embroidery, perceiving that our interconnectedness isn't simply a quality yet a core value for an agreeable presence.

6. SUSTAINABLE PRACTICES IN LIVESTOCK FARMING

ENVIRONMENTAL IMPACT OF LIVESTOCK FARMING

Livestock farming is a foundation of our worldwide food supply, yet its ecological impression poses a potential threat. As we dig into the complex trap of supportability, understanding the significant effects of this industry on our planet becomes pivotal.

Water Shortage and Contamination
Water, the remedy of life, faces uncommon strain because of animals cultivating. Picture the huge territories of farmland, where water runs scant as it sustains the thirst of millions of steers and poultry. In districts previously wrestling with water shortage, this request strengthens, frequently prompting exhaustion

of springs and stressed water supplies for nearby networks.

Additionally, the contamination brought about by escalated domesticated animals tasks is faltering. Squander overflow from ranches, loaded down with anti-microbials, chemicals, and microorganisms, tracks down its direction into streams and lakes, harming sea-going environments and endangering human wellbeing downstream. The fragile equilibrium of sea-going life experiences unsalvageable damage, setting off chain responses that reverberation through the environment.

Deforestation and Natural surroundings Annihilation

In the mission to take care of blossoming populaces, timberlands succumb to the unquenchable interest for field and feed crops. Tremendous wraps of biodiverse scenes are cleared to clear a path for cows farms and monoculture estates, taking environments from incalculable species. The deficiency of these biological systems decreases biodiversity as well as worsens environmental change, as woodlands assume a basic part in sequestering carbon dioxide.

Environmental Change

Discussing environmental change, animals cultivating remains as a huge supporter of ozone harming substance outflows. Methane, a strong ozone harming substance produced through intestinal maturation in ruminant creatures, and nitrous oxide from treated soils used to develop creature feed, further worsen an Earth-wide temperature boost. The carbon impression of delivering meat and dairy items is unquestionable, with gauges crediting a significant part of anthropogenic ozone depleting substance outflows to domesticated animals cultivating.

Soil Corruption

Underneath our feet lies one more setback from impractical cultivating rehearses: soil. Escalated brushing and monoculture crops drain soil supplements, prompting disintegration and desertification. The deficiency of fruitful soil risks the actual underpinning of horticulture, compromising our capacity to reasonably deliver nourishment for people in the future.

Embracing Maintainable Arrangements

While the difficulties presented by domesticated animals cultivating are overwhelming, trust lives in embracing reasonable practices:

Grass-took care of and Field Raised Systems: Empowering strategies that permit creatures to eat on normal field decreases dependence on grain-escalated feed frameworks and advances better environments.

Proficient Water Use: Carrying out water-effective advancements and practices can fundamentally diminish the water impression of animals cultivating.

Regenerative Agriculture: By coordinating animals into assorted crop pivots and reestablishing debased lands, regenerative farming holds guarantee in reestablishing soil wellbeing and moderating environment influences.

Lessening Food Waste: Tending to failures in food conveyance and utilization can lighten tension ashore and water assets utilized in animals creation.

STRATEGIES FOR WASTE MANAGEMENT AND RESOURCE CONSERVATION

In the domain of reasonable practices in livestock farming, one of the most pivotal viewpoints is powerful waste administration and asset protection. This guarantees ecological stewardship as well as advances productivity and financial practicality for ranchers. How about we dig into a few convincing procedures that moderate natural effect as well as improve the general maintainability of animals tasks.

1. Fertilizing the soil and Supplement Recycling:

Envision changing fertilizer and natural waste into important manure that improves soil fruitfulness. This training diminishes squander amassing as well as limits the requirement for compound manures, accordingly advancing soil wellbeing and harvest efficiency. Ranchers can coordinate treating the soil into their everyday schedules, transforming what was once viewed as waste into an important asset.

2. Biogas Production:

Outfitting methane from animals compost through anaerobic assimilation lessens ozone harming substance discharges as well as creates sustainable power. Biogas can be utilized for warming, power age, or even as a substitute for petroleum derivatives, consequently lessening reliance on non-sustainable assets and bringing down functional expenses.

3. Water Preservation Techniques:

Water is a valuable asset in cultivating, and domesticated animals tasks can execute different procedures to limit water use. Strategies, for example, water collecting, effective water system frameworks, and appropriate water reusing can essentially lessen water utilization while guaranteeing satisfactory hydration for creatures and yields the same.

4. Coordinated Nuisance The board (IPM):

By taking on IPM procedures, ranchers can decrease dependence on compound pesticides and herbicides. This approach includes utilizing normal hunters, crop pivot, and organic controls to really oversee irritations and weeds. In addition to the fact that IPM protects biodiversity and pollinators, however it likewise keeps destructive synthetic compounds from polluting soil and water assets.

5. Accuracy Taking care of and Nutrition:

Advancing animals consumes less calories through accuracy taking care of works on creature wellbeing and efficiency as well as diminishes feed waste and supplement discharge. Via cautiously forming slims down in light of wholesome requirements and occasional varieties, ranchers can limit natural effect while boosting asset productivity.

6. Maintainable Foundation and Design:

From energy-effective structures to very much planned compost storage spaces, putting resources into supportable framework can have long haul benefits. Appropriately planned offices limit spillover and scent issues, upgrade specialist security, and further develop generally speaking homestead effectiveness. Consolidating sustainable power sources like sunlight based chargers or wind turbines further diminishes natural impression and functional expenses.

7. Instruction and Training:

Engaging ranchers with information about supportable practices and advances is fundamental for far and wide reception. Studios, preparing projects, and shared learning organizations can work with the trading of thoughts and best works on, cultivating a local area focused on reasonable horticulture.

8. Joint effort and Partnerships:

Teaming up with nearby networks, ecological associations, and legislative organizations can enhance endeavors towards feasible waste administration and asset protection. Aggregate drives can prompt shared assets, financing open doors, and strategy support that benefit the two ranchers and the climate.

Carrying out viable waste management and conservation methodologies in animals cultivating isn't simply a natural obligation however a pathway to flexibility and productivity. By taking on these practices, ranchers can lessen ecological effect, improve asset productivity, and add to a more feasible future for a long time into the future. Together, we can change difficulties into open doors, guaranteeing that horiculture stays a foundation of maintainable turn of events.

CARBON FOOTPRINT REDUCTION AND RENEWABLE ENERGY USE

In the domain of supportable horiculture, tending to the carbon impression of domesticated animals cultivating is both a squeezing challenge and a promising an open door. Animals cultivating, while fundamental for food creation, is a critical supporter of ozone harming substance discharges internationally. Notwithstanding, by taking on rehearses that lessen carbon impression and embrace environmentally friendly power sources, ranchers can assume a vital part in relieving environmental change while guaranteeing the drawn out reasonability of their tasks.

Figuring out the Carbon Impression

Domesticated animals cultivating creates fossil fuel byproducts basically through methane delivered by creatures during absorption and using petroleum products for apparatus, warming, and power. These emanations add to a dangerous atmospheric devation and natural debasement. Notwithstanding, there are pragmatic advances that can fundamentally lessen this effect.

Useful Moves toward Decrease Carbon Impression

1. Further developed Domesticated animals Management:

Feed Efficiency: Advancing feed organization and taking care of practices can diminish methane emanations from animals.

Excrement Management: Executing successful fertilizer the board procedures, for example, treating the soil or methane catch, can relieve emanations and produce environmentally friendly power.

2. Sustainable power Integration:

Sunlight based Power: Introducing sunlight powered chargers on ranch structures can give sustainable power to warming, cooling, and power needs.

Wind Power: Using wind turbines on reasonable ranch land can create clean power, decreasing dependence on petroleum derivatives.

3. Supportable Land Use Practices:

Agroforestry: Establishing trees on ranch land sequesters carbon dioxide as well as turns out extra revenue through lumber and organic product creation.

Meadow Management: Rotational munching and field the executives practices can upgrade soil wellbeing, sequester carbon, and further develop biodiversity.

Monetary and Natural Advantages

Embracing these practices lessens fossil fuel byproducts as well as offers financial advantages:

Cost Savings: Environmentally friendly power sources can prompt decreased energy bills after some time.

Market Access: Buyers progressively incline toward items with a lower ecological effect, giving business sector benefits.

Administrative Compliance: Meeting carbon decrease targets might situate cultivates well in future administrative conditions.

Contextual analyses: Genuine Applications

1. Family-Possessed Dairy Farm:

By introducing sun powered chargers on their horse shelter rooftops, this homestead diminished power costs by 30% every year while altogether bringing down their carbon impression. The abundance energy produced is taken care of once again into the network, turning out an extra revenue stream.

2. Hamburger Dairy cattle Ranch:

Taking on rotational touching practices further developed field quality and creature wellbeing as well as sequestered carbon in the dirt. This farm presently fills in as a model of maintainable land the board in the locale.

Lessening the carbon impression of domesticated animals cultivating through environmentally friendly power use and supportable practices isn't just achievable yet basic for the fate of agribusiness. By taking on these systems, ranchers can upgrade their versatility to environmental change, further develop benefit, and add to a better planet for people in the future. Every little step towards manageability counts, and all in all, they can prompt significant positive effects on both nearby environments and the worldwide environment.

Allow us to hold hands in encouraging a future where reasonable practices are the foundation of animals cultivating, guaranteeing food security without compromising natural respectability.

7. TRANSPARENCY AND CONSUMER TRUST

COMMUNICATING ETHICAL PRACTICES TO CONSUMERS

In today's interconnected world, consumers are increasingly conscientious about where they invest their trust and their money. Ethical practices have emerged as a cornerstone of consumer decision-making, influencing perceptions of brands and products far beyond traditional metrics of quality and price. At the heart of this paradigm shift lies transparency; the open, honest communication of how businesses operate and the values they uphold.

The Importance of Ethical Communication

Ethical communication isn't merely a legal or moral obligation; it is a strategic imperative for businesses aiming to cultivate enduring relationships with their customers. When consumers are informed about a company's ethical practices, they feel empowered to make choices aligned with their own values. This empowerment fosters loyalty and advocacy, transforming satisfied customers into brand ambassadors.

Strategies for Effective Communication

1. Clear and Accessible Information: Transparency begins with clarity. Businesses should ensure that information about their ethical standards and practices is readily accessible. This includes prominently displaying policies on their websites, in-store signage, and packaging.

2. Storytelling with Impact: Beyond facts and figures, storytelling humanizes ethical practices. Sharing stories of how ethical decisions positively impact stakeholders;

from employees to local communities; resonates with consumers on an emotional level.

3. Engagement and Dialogue: Authenticity is key to building trust. Engage with consumers through social media, community events, or feedback mechanisms. Actively listening to concerns and responding transparently demonstrates commitment to ethical values.

4. Certifications and Third-Party Validation: External validation through certifications from reputable organizations adds credibility. These endorsements assure consumers that ethical claims are backed by independent assessment.

Real-Life Applications

Consider a local farm-to-table restaurant that prides itself on sustainability. By transparently communicating its commitment to sourcing organic produce and supporting local farmers, it not only attracts environmentally conscious diners but also educates them about responsible dining choices.

Similarly, a fashion brand that embraces fair trade practices and provides detailed insights into its supply chain empowers consumers to make informed decisions about their purchases. This transparency builds a loyal customer base that values ethical manufacturing and worker welfare.

Challenges and Overcoming Them

While the benefits of ethical communication are clear, challenges such as greenwashing (misleading environmental claims) or insufficient disclosure remain. Businesses must guard against these pitfalls by adhering to rigorous standards and being accountable for their claims.

Ethical communication is not just a trend but a fundamental shift towards a more sustainable and conscientious marketplace. By embracing transparency and effectively communicating their ethical practices, businesses not only comply with regulatory requirements but also inspire consumer confidence, foster loyalty, and contribute positively to society. As consumers continue to prioritize values alongside products, ethical communication becomes the cornerstone of building lasting trust and driving meaningful change in the marketplace.

LABELING AND CERTIFICATION PROGRAMS

In the present complex commercial center, shoppers are immersed with decisions going from ordinary items to specific administrations. In the midst of this overflow, how could you, as an upright customer, settle on informed choices that line up with your qualities and needs? This is where marking and certificate programs move toward, offering a signal of clearness in an ocean of choices.

Understanding Marking and Affirmation Projects

Naming and certificate programs are instruments intended to give straightforward data about items and administrations. They engage shoppers by offering experiences into different viewpoints like fixings, obtaining strategies, natural effect, and moral practices. These projects go about as an extension among makers and purchasers, encouraging trust through transparency and responsibility.

Why Straightforwardness Matters

Envision strolling into a supermarket, searching for natural produce or morally obtained espresso. The variety of decisions can be overpowering without obvious signs of what lines up with your inclinations. Marking and certificate programs slice through this disarray. They offer confirmation that the cases made by brands are checked by free bodies, guaranteeing validity and dependability.

Benefits Past the Mark

Past improving on buyer decisions, these projects drive positive change across enterprises. By setting guidelines and empowering consistence, they advance manageable practices, fair work conditions, and capable obtaining. This advantages purchasers as well as supports organizations focused on moral lead, making an expanding influence of social and ecological stewardship.

Enabling Buyer Decision

Consider the effect of realizing that your buy upholds fair exchange standards, safeguards jeopardized species, or advances environmentally friendly power. Marking and certificate programs engage you to cast a ballot with your wallet, impacting ventures towards more noteworthy manageability and moral obligation. Your decisions become a strong impetus for change, intensifying the interest for items that focus on both quality and trustworthiness.

Building Trust and Responsibility

Trust is the foundation of any relationship, including that among shoppers and brands. Marking and affirmation programs develop trust by considering organizations responsible for their cases. Through thorough assessment and consistence observing, these projects guarantee that what is guaranteed is conveyed, encouraging long haul customer faithfulness and brand notoriety.

Certifiable Applications

Picture a reality where each item on the rack accompanies a reasonable mark showing its natural effect, healthy benefit, and moral contemplations. This straightforwardness illuminates your dynamic cycle as well as interfaces you with the upsides of the organizations behind the items. Whether you're picking beauty care products, gadgets, or family basics, naming and accreditation programs give the straightforwardness expected to pursue decisions that resound with your convictions.

The Job of Advancement and Versatility

As buyer inclinations advance and worldwide difficulties, for example, environmental change increase, marking and affirmation programs should keep on enhancing. They should adjust to resolve arising issues, for example, carbon impressions or store network straightforwardness, guaranteeing importance and viability in a powerful commercial center.

Marking and confirmation programs are not just about stickers on items; they address a pledge to straightforwardness, honesty, and informed purchaser decision. By supporting these projects, you add to a more manageable and dependable worldwide economy. Together, we can fabricate a future where straightforwardness and purchaser trust are the standard, enabling people to have a constructive outcome through regular buying choices.

BUILDING TRUST THROUGH ACCOUNTABILITY AND TRACEABILITY

In the present interconnected world, where decisions flourish and data is promptly available, trust has turned into the foundation of effective connections among organizations and customers. The substance of trust lies in straightforwardness, and the way to accomplishing straightforwardness lies in responsibility and recognizability.

Envision you're looking for food, and you run over two brands of natural produce. Both case to be sans pesticide and harmless to the ecosystem. Be that as it may, one brand goes above and beyond; it gives point by point data about the homesteads where the produce was developed, the techniques utilized for development, and even offers endorsements from autonomous examiners checking their cases. The other brand, then again, offers no such straightforwardness; their bundling just states "natural" with no extra subtleties.

The greater part of us would incline towards the primary brand since it shows responsibility and detectability. This straightforwardness guarantees us of the item's validness as well as lines up with our upsides of maintainability and moral utilization. By enthusiastically sharing data about their cycles and welcoming examination from outsider reviewers, the primary brand constructs a relationship of trust with its clients. They comprehend that trust is procured through transparency and genuineness, not simply through shrewd promoting strategies.

Similar standards apply across different businesses, from innovation to medical services and then some. Consider a product organization that creates applications dealing with touchy individual information. In the present environment of information breaks and security concerns, clients are naturally mindful about where their data goes. An organization that focuses on responsibility and detectability won't just follow information security guidelines yet will likewise proactively illuminate clients about how their information is gathered, put away, and utilized. They could offer apparatuses for clients to deal with their protection settings and furnish ordinary reviews to guarantee consistence with security norms.

Additionally, in medical care, trust is principal. Patients depend on clinical experts and drug organizations to give precise data about medicines, meds, and likely aftereffects. A drug organization that embraces straightforwardness through responsibility and discernibility will lead thorough testing, share point by point clinical preliminary outcomes, and uncover any unfriendly occasions related with their items. This degree of straightforwardness not just engages patients to settle on informed conclusions about their wellbeing yet additionally encourages a relationship of trust between medical care suppliers and the networks they serve.

Basically, building trust through responsibility and discernibility isn't simply a business technique; it's a guarantee to respectability and moral obligation. At the point when organizations make their ways for examination, they exhibit trust in their practices and regard for their clients' on the right track to be aware. This approach draws in faithful clients as well as develops a positive standing that can endure difficulties and emergencies.

As customers, we have the ability to request straightforwardness from the brands we support. By deciding to belittle organizations that focus on responsibility and detectability, we send a reasonable message that uprightness matters. Together, we can make a commercial center where trust isn't simply a trendy expression yet a basic rule that directs our collaborations and shapes the eventual fate of business.

Whether you're purchasing food, downloading an application, or arriving at conclusions about your wellbeing, recall that straightforwardness assembles trust. Pick shrewdly, support responsibility, and request detectability; in light of the fact that in reality as we know it where data is top dog, trust is the cash that makes the biggest difference.

8. INNOVATIONS AND ADVANCES IN ETHICAL FARMING

TECHNOLOGICAL INNOVATIONS IN LIVESTOCK MANAGEMENT

In the domain of moral cultivating, where the prosperity of creatures is principal, mechanical developments have arisen as an encouraging sign and progress. These progressions improve the proficiency of animals the executives as well as hoist the guidelines of creature government assistance to remarkable levels.

1. Accuracy Animals Farming:

Gone are the times of summed up care for domesticated animals. Accuracy animals farming (PLF) uses sensors and information examination to screen individual creatures' wellbeing and conduct progressively. Envision a dairy ranch where each cow's milk creation, taking care of examples, and even development are fastidiously followed. This innovation empowers ranchers to recognize early indications of disease, upgrade taking care of systems, and make customized care plans, accordingly limiting pressure and amplifying the general soundness of the group.

2. Robotized Wellbeing Observing Systems:

In customary cultivating, distinguishing a debilitated creature frequently depended on obvious prompts, which some of the time implied sicknesses were just identified at cutting edge stages. With robotized wellbeing checking frameworks, for example, wearable gadgets outfitted with biosensors, ranchers get moment cautions in the event that an animal's important bodily functions veer off from the standard. This proactive methodology further develops treatment results as well as decreases the requirement for intrusive mediations.

3. Mechanical technology and man-made intelligence Helped Feeding:

Mechanical frameworks have upset taking care of practices on ranches by definitively apportioning feed in light of dietary necessities determined through artificial intelligence calculations. This limits food wastage as well as guarantees that every creature gets a fair eating regimen custom-made to its particular requirements. Also, automated milkers have changed the dairy business, offering cows the opportunity to be drained on their own timetable, in this way decreasing pressure and improving in general efficiency.

4. Ecological Observing and Control:

Animals cultivating frequently converges with ecological difficulties. High level observing frameworks presently permit ranchers to follow air quality, moistness levels, and temperature varieties inside stables or poultry houses. Mechanized ventilation and environment control frameworks change continuously to make ideal circumstances for creature solace and wellbeing. Such advancements work on animal government assistance as well as add to economical cultivating rehearses by limiting natural effect.

5. Information Driven Choice Making:

The coordination of enormous information examination in animals the board empowers ranchers to pursue informed choices in light of extensive bits of knowledge. Authentic information on reproducing designs, hereditary profiles, and development directions enable ranchers to advance rearing projects, foresee market requests, and decisively plan ranch tasks. This information driven approach improves efficiency as well as cultivates a more manageable and moral cultivating climate.

Mechanical developments in animals the board are not only apparatuses of proficiency but rather foundations of moral cultivating rehearses. By focusing on the prosperity of animals through accuracy, computerization, and information driven bits of knowledge, these developments prepare for a future where cultivating is both feasible and others conscious. As we keep on embracing these headways, we maintain our obligation to mindful stewardship of domesticated animals and guarantee a more promising time to come for a long time into the future.

Fundamentally, the excursion towards moral cultivating is directed by these groundbreaking advances, where sympathy meets development to reclassify the principles of domesticated animals the executives in the 21st 100 years.

RESEARCH AND DEVELOPMENT IN ANIMAL WELFARE SCIENCE

In the domain of moral cultivating, the wilderness of innovative work in creature government assistance science remains as an encouraging sign and progress. It's not just about satisfying industry guidelines; it's tied in with rethinking them, lifting our ethical compass, and guaranteeing a maintainable future for the two creatures and people the same.

Envision an existence where each cow, pig, or chicken carries on with a day to day existence, however an existence of nobility and solace. This vision isn't idealistic; it's attainable through committed innovative work in creature government assistance science. These endeavors dive profound into grasping creature conduct, physiology, and brain science to tailor conditions and practices that focus on their prosperity.

Take, for example, the spearheading work in grasping pressure reactions in domesticated animals. By fastidiously concentrating on cortisol levels and conduct prompts, scientists have revealed nuanced bits of knowledge into how natural variables; from temperature and lighting to social elements, influence creatures. Outfitted with this information, ranchers can now advance everyday environments, lessening feelings of anxiety and further developing in general wellbeing results.

In addition, progressions in sustenance science have prompted customized slims down that improve creature wellbeing and decrease natural effect. From adjusted micronutrient details to maintainable feed obtaining, each viewpoint is examined to guarantee ideal nourishment without compromising moral guidelines.

Be that as it may, maybe most convincing are the leap forwards in veterinary consideration and sickness counteraction. Envision a situation where illnesses are distinguished right on time through state of the art diagnostics and treated immediately with insignificant disturbance to the creature's daily practice. This proactive methodology works on creature government assistance as well as protections food security by guaranteeing sound domesticated animals populaces.

Past the ranch entryways, research in animal government assistance science reaches out to strategy promotion and shopper training. Moral cultivating rehearses are not only a trendy expression; they are established in proof based science that illuminates guidelines and engages buyers to settle on informed decisions. Straightforwardness in cultivating rehearses constructs trust and encourages a temperate cycle where moral norms are maintained from ranch to table.

In useful terms, these advancements convert into substantial enhancements. Ranches furnished with robotized observing frameworks can right away change natural boundaries to guarantee ideal solace for animals. Developments like these upgrade efficiency as well as cultivate an agreeable connection between creature government assistance and horticultural productivity.

Eventually, innovative work in animal government assistance science are not just about pushing limits; they are tied in with revising the account of moral cultivating. It's tied in with recognizing our obligation as stewards of the Earth and resolving to rehearses that regard the characteristic worth of each and every living animal.

As we look forward, the excursion towards moral cultivating will keep on being directed by logical request and moral contemplations. Together, through cooperative endeavors between scientists, ranchers, policymakers, and shoppers, we can make a future where practical farming flourishes close by sympathy for animals; a future where development in animal government assistance science isn't simply a decision yet an ethical objective.

FUTURE TRENDS AND DIRECTIONS IN ETHICAL FARMING

Developments and Advances in Moral Cultivating have introduced another time where supportability, creature government assistance, and moral practices are at the very front of horticultural practices. Looking forward, the direction of Moral Cultivating guarantees significantly more extraordinary changes that are essential as well as basic for the prosperity of our planet and its occupants.

1. Maintainable Farming Practices:

The eventual fate of Moral Cultivating depends on manageability. Practices like regenerative agribusiness, natural cultivating, and permaculture are picking up speed as they focus on soil wellbeing, biodiversity, and normal asset protection. Ranchers are progressively embracing coordinated bug the executives procedures, crop turn, and water-effective water system frameworks to limit ecological effect while guaranteeing long haul efficiency.

2. Innovation and Innovation:

Headways in innovation are set to alter Moral Cultivating. From accuracy cultivating utilizing robots and satellite symbolism to artificial intelligence driven investigation for advancing harvest yields, innovation is making farming more effective and supportable. Developments like vertical cultivating and tank-farming are considering food creation in metropolitan regions, decreasing the carbon impression related with transportation and capacity.

3. Straightforwardness and Purchaser Awareness:

Moral Cultivating is turning out to be more straightforward, driven by purchaser interest for discernibility and responsibility. Names like natural, fair exchange, and mercilessness free are patterns as well as impressions of a developing moral cognizance among purchasers. This pattern is pushing ranchers and food makers to embrace more moral practices, from altruistic treatment of animals to fair work rehearses across the production network.

4. Creature Government assistance and Moral Treatment:

The fate of Moral Cultivating additionally envelops critical steps in creature government assistance. There is a shift towards giving creatures better everyday environments, admittance to outside spaces, and normal ways of behaving. Innovations, for example, wearable wellbeing screens for animals and high level lodging frameworks are being created to guarantee that creatures are dealt with empathetically all through their lives.

5. Worldwide Coordinated effort and Strategy Development:

Tending to moral worries in cultivating requires worldwide joint effort and powerful approach structures. Worldwide associations, states, and NGOs are progressively cooperating to set norms for Moral Cultivating rehearses, advance fair exchange, and uphold guidelines that safeguard the two creatures and the climate. These endeavors are pivotal in making a level battleground where moral practices are compensated and untrustworthy practices are gotten rid of.

6. Environmental Change Adaptation:

As environmental change presents difficulties to horticulture, Moral Cultivating will assume an essential part in variation and relief systems. Ranchers are investigating dry spell safe harvests, carbon sequestration techniques, and strong horticultural practices to shield food security in an evolving environment. Moral contemplations in cultivating will become entwined with endeavors to battle environmental change, guaranteeing that horticulture stays practical and strong.

Taking everything into account, the eventual fate of Moral Cultivating is promising yet testing. It requires consistent development, joint effort, and obligation to moral standards. As purchasers become more mindful and requesting, and as innovation keeps on advancing

9. CASE STUDIES AND BEST PRACTICES

SUCCESSFUL MODELS OF ETHICAL FARMING OPERATIONS

In the domain of horticulture, the idea of moral cultivating has risen above simple practice to turn into a guide of manageability and obligation. As we dive into contextual analyses and best works on, investigating fruitful models of moral cultivating activities reveals insight into powerful systems as well as highlights their significant effect on environments, networks, and worldwide food security.

Embracing Regenerative Agriculture:
One commendable model rises up out of the core of the Midwest, where a family-claimed ranch has spearheaded regenerative farming. By fastidiously turning crops, incorporating

domesticated animals, and rehearsing negligible culturing, they have revived soil wellbeing and biodiversity. The outcomes say a lot: expanded yields, diminished dependence on engineered inputs, and a quantifiable reduction in carbon impression. This approach guarantees long haul efficiency as well as encourages versatility against environmental change, typifying an agreeable connection among nature and cultivating.

Local area Focused Practices:

In the moving slopes of Vermont, another ranch has embraced a local area focused way to deal with moral cultivating. Through direct-to-shopper models like CSA and rancher's business sectors, they have manufactured profound associations with nearby inhabitants. This not just ensures a steady pay for the ranch yet additionally advances food power and teaches customers on the worth of supportable practices. By focusing on straightforwardness and joint effort, they have developed a steadfast client base put resources into the ranch's moral process.

Creature Government assistance Excellence:
Across the fields of Australia, a farm remains as a guide of moral animal cultivation. By focusing on the government assistance of their domesticated animals, they have embraced field based frameworks that emulate normal natural surroundings. Steers munch uninhibitedly on open prairies, guaranteeing they lead tranquil lives and produce unrivaled quality meat. This obligation to moral treatment not just lines up with customer inclinations for sympathetic practices yet in addition highlights the homestead's devotion to capable stewardship of assets.

Imaginative Innovation Integration:

In the lavish fields of the Netherlands, development meets custom in a homestead that has consistently coordinated innovation into moral cultivating rehearses. Robotized frameworks screen soil dampness levels, advance water system, and limit squander, all while decreasing work force. By tackling the force of information investigation and accuracy agribusiness, they have accomplished higher efficiencies without compromising ecological trustworthiness. This model sets benchmarks for efficiency as well as motivates others to embrace state of the art arrangements in quest for feasible cultivating.

Instructive Drives and Information Sharing:

In the lower regions of India, a helpful of smallholder ranchers has flourished through instructive drives and information sharing. By engaging ranchers with preparing in natural cultivating strategies and economical bug the executives, they have improved efficiency while safeguarding biodiversity. This grassroots methodology elevates provincial networks as well as makes an expanding influence of positive change across ages. Through studios and rancher field schools, they cultivate a culture of consistent learning and variation, guaranteeing that moral cultivating rehearses persevere.

Fruitful models of moral cultivating tasks are not just about returns or benefits; they epitomize a comprehensive obligation to the planet and its occupants. They show the way that benefit and manageability can coincide amicably, demonstrating that capable stewardship of normal assets isn't simply a decision however an ethical objective. As we keep on investigating contextual analyses and best practices in moral cultivating, let us draw motivation from these commendable models and on the whole endeavor towards a future where farming feeds the two individuals and the planet.

In embracing these models, we plant the seeds of a supportable rural transformation; one that guarantees a plentiful reap for a long time into the future.

LESSONS LEARNED FROM GLOBAL PERSPECTIVES

Our general surroundings is quickly developing, and the capacity to adjust to these progressions is vital for progress. To flourish in the present worldwide society, it's fundamental to gain from the encounters of others and expand one's perspectives. This is the reason looking at worldwide points of view is more significant now than any other time in recent memory. Through various encounters, difficulties, and victories, we can foster a superior comprehension of the world we live in and the examples it needs to educate us.

One illustration we can gain according to worldwide points of view is the significance of embracing variety. As we interface with others from various societies, we open ourselves to novel thoughts and perspectives. At the point when we embrace this mentality, we are better prepared to coincide and make a more comprehensive world. This is exemplified by the narrative of Arundhati Roy, a famous Indian creator who straightforwardly advocates for minimized networks and the individuals who have been hushed. Her all consuming purpose fills in as a sign of how variety, compassion, and inclusivity can decidedly affect our reality.

Another significant illustration we gain according to looking at worldwide viewpoints is the significance of coordinated effort. At the point when we work with others and offer our thoughts, we can possibly accomplish incredible things. The Unified Countries' Supportable Improvement Objectives (SDGs) give a great representation of this. These objectives were made to urge worldwide cooperation to address a portion of the world's most major problems, like destitution, imbalance, and ecological debasement. Teaming up on a worldwide scale gives different points of view and imaginative answers for issues we may in all likelihood never have thought about in any case.

Ultimately, looking at worldwide viewpoints shows us the significance of flexibility. Since the beginning of time, we have confronted various difficulties, from catastrophic events to pandemics. Notwithstanding, by persisting and adjusting to evolving conditions, we can conquer these difficulties and develop further. A fantastic illustration of this is the flexibility shown by individuals of Haiti, who have encountered monetary, political, and ecological battles from the beginning of time. Regardless of these difficulties, they proceed to reconstruct and flourish through local area based drives and projects.

Analyzing worldwide points of view can possibly show us fundamental examples in variety, cooperation, and strength. The world is continually changing, and by gaining from the encounters of others, we can more readily explore these changes. Variety and inclusivity, cooperation, and flexibility are essential attributes that will guarantee our progress in the years to come. By applying these illustrations to our regular routines, we can make a superior world for us and for people in the future.

CHALLENGES AND OPPORTUNITIES FOR IMPROVEMENT

In any work or industry, there will continuously be difficulties. Be that as it may, with challenges come open doors for development. This is particularly evident in the field of medical care, where difficulties and potential open doors meet consistently, influencing patients and experts the same.

Quite possibly of the greatest test confronting medical care today is the lack of medical services laborers. This is particularly obvious in specific strengths, like nursing and essential consideration. In any case, this challenge additionally presents an open door, by putting resources into schooling and preparing programs for medical services laborers, we can assist with easing the deficiency and guarantee that patients approach the consideration they need.

One more test in medical services is the always inflating expenses of clinical consideration. This overburdens patients, who might battle to pay for important medicines, as well as on medical care suppliers, who might battle to give quality consideration inside financial plan limitations. Once more, nonetheless, this challenge presents an open door; by executing cost-saving measures, like deterrent consideration and telemedicine, we can assist with decreasing expenses without forfeiting quality consideration for patients.

Quite possibly of the most squeezing challenge in medical care today is the continuous Coronavirus pandemic. This has overwhelmed medical services laborers and frameworks across the globe, and has featured the requirement for further developed contamination control measures and crisis readiness. In any case, even amidst this test, there have been open doors for development, medical services suppliers have quickly adjusted to new strategies for care conveyance, for example, telehealth arrangements, to guard patients.

While there are surely difficulties confronting medical care today, there are likewise various open doors for development. By putting resources into schooling and preparing programs, carrying out cost-saving measures, and adjusting to new strategies for care conveyance, we can assist with guaranteeing that patients get the most ideal consideration, even notwithstanding obstructions. It is our obligation as medical services suppliers and supporters to quickly jump all over these chances for development and keep on pushing for progress in the field of medical services.

10. CONCLUSION AND FUTURE OUTLOOK

SUMMARY OF KEY FINDINGS AND INSIGHTS

In investigating the many-sided scene of our exploration, we have uncovered significant bits of knowledge that enlighten our way ahead. Our excursion through information and examination has uncovered patterns and examples as well as given clearness on vital issues that shape how we might interpret topic.

Revealing Examples and Trends

One of the focal discoveries of our examination is the rise of unmistakable examples inside field/industry. Through fastidious information assortment and thorough examination, we have recognized patterns that feature shifts in purchaser conduct, market elements, and mechanical headways. These patterns highlight the present status of undertakings as well as proposition significant premonition into future turns of events.

Grasping Customer Behavior

Digging further into buyer conduct, our exploration has uncovered intriguing bits of knowledge into the inspirations and inclinations driving dynamic cycles. By looking at purchaser criticism and overview information, we have acquired a nuanced comprehension of what drives consumer loyalty and steadfastness in market/industry. These bits of knowledge are critical for making methodologies that resound with our interest group and upgrade client commitment.

Suggestions for Technique and Innovation

The ramifications of our discoveries stretch out past simple perception; they act as a compass directing vital choices and cultivating development. By utilizing our bits of knowledge, associations can recalibrate their ways to deal with item advancement, advertising efforts, and client assistance drives. This proactive position mitigates takes a chance as well as positions us at the front line of industry patterns, ready for reasonable development and upper hand.

Challenges and Opportunities

However, our process has additionally enlightened difficulties that warrant consideration. From administrative obstacles to mechanical interruptions, these difficulties present open doors for innovativeness and strength. By tending to these obstructions head-on, we can change affliction into a potential open door, driving significant change and laying out new benchmarks of greatness in our field.

Looking Forward: Vision and Future Outlook

As we close this period of investigation, our vision for what's in store is clear and convincing. Equipped with powerful bits of knowledge and a relentless obligation to development, we are ready to spearhead new arrangements, fashion key organizations, and lift principles of greatness in [industry/sector]. Our process doesn't end here; rather, it develops into a promising future where our discoveries act as an impetus for development, variation, and change.

Our investigation into the vital discoveries and experiences of topic has been edifying and extraordinary. By embracing these experiences, we engage ourselves to explore intricacies with certainty, drive significant change, and shape a future that isn't just manageable yet in addition significantly effective. Together, let us set out on this excursion of revelation, furnished with information and driven by a common obligation to greatness and development.

RECOMMENDATIONS FOR POLICY MAKERS AND STAKEHOLDERS

In charting the course for future progress and sustainable development, the role of policy makers and stakeholders cannot be overstated. As we navigate the complexities of a rapidly changing world, it becomes imperative to adopt strategies that not only address current challenges but also pave the way for a resilient and inclusive future. Here are compelling recommendations tailored to empower policy makers and stakeholders in their crucial roles:

Embrace Collaborative Governance:

Effective policy making requires collaboration across sectors and disciplines. Stakeholders from government, industry, academia, and civil society must come together to co-create solutions that are both innovative and inclusive. By fostering dialogue and partnership, policy makers can leverage diverse perspectives and expertise to tackle complex issues such as climate change, healthcare access, and economic inequality.

Prioritize Long-Term Sustainability:

Short-term gains should not compromise long-term sustainability. Policy makers must adopt a forward-thinking approach that considers environmental, social, and economic impacts over time. Investing in renewable energy, promoting sustainable agriculture, and integrating circular economy principles are essential steps towards ensuring a prosperous future for generations to come.

Invest in Education and Skills Development:

Empowering individuals through education and skills development is key to unlocking human potential and fostering inclusive growth. Policy makers should prioritize investments in education systems that equip learners with the knowledge and skills needed for the jobs of tomorrow. This includes promoting STEM education, digital literacy, and vocational training programs that cater to diverse learner needs.

Foster Innovation and Digital Transformation:

In a digital age, embracing technological innovation is paramount to driving economic growth and enhancing public service delivery. Policy makers should create an enabling environment for entrepreneurship and innovation, including supportive regulatory frameworks and incentives for research and development. Embracing digital transformation can also improve access to healthcare, education, and financial services, particularly in underserved communities.

Promote Social Equity and Inclusion:

Addressing systemic inequalities requires intentional policies that promote social equity and inclusion. Policy makers should prioritize initiatives that reduce barriers to opportunity, such as affordable housing programs, healthcare access for vulnerable populations, and measures to combat discrimination in all its forms. By fostering a society where everyone can thrive, we ensure a more just and cohesive future.

Strengthen Resilience to Global Challenges:

In an interconnected world, resilience is key to mitigating the impacts of global challenges such as pandemics, climate change, and economic volatility. Policy makers should invest in robust healthcare systems, disaster preparedness strategies, and social safety nets that can withstand shocks and protect vulnerable communities. Building resilience requires foresight, adaptive planning, and a commitment to collective action.

Enhance Transparency and Accountability:

Trust in public institutions is essential for effective governance. Policy makers should prioritize transparency in decision-making processes and ensure accountability for actions taken. This includes promoting open data initiatives, engaging citizens in policy discussions, and combating corruption at all levels. By fostering a culture of transparency and accountability, policy makers can build trust and legitimacy in their governance efforts.

As we look towards the future, the decisions made by policy makers and stakeholders today will shape the world of tomorrow. By embracing collaborative governance, prioritizing sustainability, investing in education and innovation, promoting social equity, strengthening resilience, and enhancing transparency, we can create a future that is prosperous, inclusive, and sustainable for all. Together, we have the power to drive positive change and build a better world for generations to come.

THE PATH FORWARD: ENHANCING LIVESTOCK WELFARE AND ETHICAL FARMING

In imagining the fate of agriculture, a urgent perspective that requests our prompt consideration is the government assistance of domesticated animals and the advancement of moral cultivating rehearses. This isn't just about industry guidelines; it's about our ethical obligation to guarantee that each living being under our consideration carries on with a daily existence deserving of regard and respect.

Envision a reality where each livestock encounters benevolence from birth to the furthest limit of their lives. This vision isn't idealistic; it's reachable through coordinated endeavors and an aggregate change in outlook towards moral cultivating. This is the way we can clear the way ahead:

1. Bringing issues to light and Education:

Schooling is the foundation of progress. By bringing issues to light among ranchers, shoppers, and policymakers about the significance of animals government assistance, we can drive an interest for moral practices. Studios, classes, and instructive missions can assume a significant part in cultivating this comprehension.

2. Carrying out Tough Standards:

Guidelines and principles should develop to focus on creature government assistance. From guaranteeing sufficient day to day environments to compassionate butcher practices, each step of the production network ought to mirror our obligation to moral treatment. State run administrations and industry bodies can team up to authorize and persistently work on these principles.

3. Supporting Reasonable Cultivating Methods:

Moral cultivating isn't just about creature government assistance; it's likewise about maintainability. Empowering rehearses like rotational brushing, natural cultivating, and diminished anti-microbial utilize benefit creature wellbeing as well as add to ecological protection. Purchasers can uphold these endeavors by picking items that stick to such standards.

4. Putting resources into Innovation:

Innovative headways offer promising arrangements. From simulated intelligence driven observing frameworks that distinguish indications of misery in animals to manageable feed options, development can improve both animal government assistance and ranch productivity. Putting resources into innovative work in this field is vital for constant improvement.

5. Teaming up for Change:

Accomplishing inescapable reception of moral cultivating rehearses requires joint effort across areas. Ranchers, veterinarians, specialists, and policymakers should work connected at the hip to conquer difficulties and offer accepted procedures. Cooperative endeavors can prompt useful arrangements that benefit the two creatures and the farming business overall.

6. Enabling Buyer Choice:

Shoppers employ huge impact through their buying choices. By selecting items affirmed for moral treatment of creatures, purchasers can drive market interest towards additional compassionate practices. Straightforwardness in marking and accreditation plans can enable purchasers to pursue informed decisions lined up with their qualities.

7. Observing Achievement Stories:

Featuring effective contextual investigations and accounts of homesteads that have embraced moral practices can move others to go with the same pattern. These accounts feature the possibility of moral cultivating as well as exhibit its advantages concerning animal prosperity, item quality, and rancher fulfillment.

www.ingramcontent.com/pod-product-compliance
Lightning Source LLC
Chambersburg PA
CBHW071925210526
45479CB00002B/557